Index

Introduction... 3

How to Choose the Right Type of Antenna................ 5

How to Choose the Right Material for your Antenna..... 11

How to Choose the Right Size Antenna.................. 15

How to Choose the Right Type of Matching for your Antenna......... 19

How to Construct your Antenna......................... 21

How to Support your Antenna........................... 25

How to Choose the Right Feeder for your Antenna....... 29

How to Match your Antenna and Feeder.................. 33

How to Match your Feeder to your Radio................ 37

How to Find Contacts with your new System............. 41

Introduction.

Why write yet another antenna book, isn't everything you need to know about antennas already written?

Well yes it is and I certainly have a library full of books and magazines on antennas of all styles, shapes and sizes. Now I've been an avid devourer of all things antenna for around 30 years, I even did a TAFE (Technical and Further Education – an adult education system in Australia) subject on antennas to learn more about my favourite subject so I have a very solid understanding of how they work and how to design and build them.

But, and this is a big but, they really come in two flavours. Too simplistic with little or no useful information and too technical for most people. The simplistic ones are usually very generalised and do little more than describe antennas of various types with perhaps a formula for building a single type of antenna by giving you the actual dimensions rather than the how.

The technical ones tell you all of the theory and formulae but little actual construction information particularly what materials to use etc.

What I have done here is written a book which provides you the means and methods to use to design and construct your own antenna for your favourite frequency which fits your space. Some of the materials you can use which you possibly already have or can get cheaply and some information on how to assess materials for use.

A book to cover all the different types of antennas would be much larger than this one so I decided to concentrate on the simpler ones to build and use, wire antennas. Specifically dipoles as horizontal or vertical antennas.

Don't discount a simple dipole as an effective antenna. Properly matched and balanced these will often outperform a commercial beam antenna as I have proved many times even when I'm using less power than the operator with the beam.

You can build these relatively quickly and easily. The listening and contacts you can enjoy with one of these can provide you with many, many hours of pleasure and the discussions you can have about your construction can add even more pleasure.

I am happy to discuss antennas with you, visit my website at **www.bearlybooks.com** and contact Mr Bearly.

How to Choose the Right Type of Antenna.

It is important to have an understanding of the performance characteristics of the different types of antenna and combine them with the limitations of your environment.

Not everyone has the space to construct the optimum antenna for every band you want to operate in so you will have to decide which band is going to be your favourite, for now at least.

It's pretty easy to make your antenna operate an several bands but it can only be optimised for one.

The band you choose will, to a certain extent, determine the physical size of your antenna as well.

So how then do we make those decisions if we are just staring out as a listener or ham?

It's a lot easier for a listener than a ham because there is no transmit requirement so different materials can be used and some antenna types can be really broad-band. I'll try to differentiate where appropriate but if I forget just check the matching type it should tell you where they work best.

A little maths, sorry but when you are designing an antenna you really do need to be able to do some simple calculations for best results.

Before I start that though a little antenna theory is in

order. The entire radio spectrum that you'll be interested in ranges from 3Mhz to 30Mhz. I'm ignoring everything outside the HF range and all the fascinating world of VLF but the maths and theory remains the same for those regions as well.

There is a direct relationship between the frequency and the wavelength of the signal. The lower the frequency the longer the wavelength and vice versa. Wavelength is defined as the length, in air, of two sections of the radio wave which are in the same phase and are at the same signal strength.

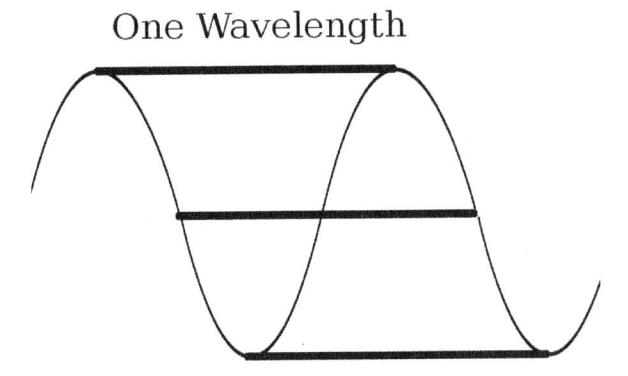

One Wavelength

For a visual image imagine the waves at the beach, the wavelength would be the distance between consecutive tops.

Radio waves are sine waves just like the waves at the beach (roughly at least). Remember sine plots from school? Yeah, I know.

Time for a little maths.

The way to calculate the wavelength for a given frequency is to divide 300 by the frequency, in Mhz.

That gives you the wavelength in metres. So, if your frequency of choice is 10Mhz 300 / 10 = 30m. For those in non-metric countries the formula is 984 divided by the frequency in Mhz to give the length in feet. While it might seem that it would be a good idea to build an full wave antenna it is actually difficult to match well and so convention has it that we use a ½ wave antenna as this is the smallest size which is resonant.

What's that? A quarter wave is resonant? Actually no it isn't, it is usually attached to radials or a solid ground plane or has a base loading section for matching all of which emulate another quarter wave antenna electrically so you see it's still a ½ wave antenna really.

I bet you have figured it out how to work out how big a ½ wave antenna is haven't you. Yep, you got it, we divide the wavelength by 2. That then provides you with the detail you need to know to work out the maximum space you'll need to operate a full-sized ½ wave antenna.

There is a bit of a proviso here, you don't really need to mount all the antenna in a straight line, I'll get to that bit later but if you want to operate on the 30m band which requires a 15m antenna but only have 6m of space and it isn't in a straight line then don't panic. I'll show you how to make it fit and work well.

At this stage I have been basically alluding to horizontal antennas (well roughly horizontal) but you can mount these antennas vertically or sloped as well which will take up a whole lot less horizontal yard space.

Apart from the space requirement, what are the differences between vertical and horizontal antennas? The main one is the angle of radiation from the antenna. This, in general, determines the distance you'll be able to hear or communicate with.

All radiation patterns are calculated in free space because trying to work out what the real pattern is with all possible reflections and radiation patterns is extremely complex so we just don't, OK?

In free space the radiation pattern around an antenna is roughly a doughnut shape at right angles to the antenna and the antenna in the centre of the hole.

Thinking about that, the horizontal antenna has a maximum radiation strength at right angles to the antenna but at a range of angles from ground level to straight up. Minimum strength is in line with the antenna.

The vertical antenna has maximum radiation strength at right angles to the antenna but the strength varies from strongest at zero degrees to the horizon and almost none straight up.

Putting up a vertical means you can receive signals from all over but a horizontal antenna can only receive signals from areas in front and behind for about a 60 degree arc.

The other issue is the height off the ground that you can mount the feed point of the antenna. They all work best between ¼ and ½ a wavelength from the ground. That ground is considered to be the actual physical area under the antenna, which could actually be the roof it's mounted on. It's complicated isn't it? Here's something to cheer you up.

It is almost impossible to build an antenna that doesn't work.

How much space do you have to put your antenna up in? See that's actually the determining factor for both the antenna and the frequency of operation. The available space sets whether the antenna type is horizontal or vertical, the frequency of operation sets the size. We just have to live with the radiation patterns.

You task, should you choose to accept it, is to choose a frequency of operation and an antenna style.

For the remainder of the book I will be using 28Mhz as my operating frequency as it is the centre frequency of the 10m phone band. This antenna will be a maximum of 300/28 = 10.7m / 2 = 5.4m long, 984/28 = 35.14'/2 = 17.5'.

If that doesn't mean much to you it just means that you are not a ham radio operator. Adjust to suit your frequency of operation.

Oh, by the way, the radiation pattern of an antenna is the same for receive as transmit.

Ummm, well that is the conventional wisdom and it holds true for all basic antenna types but I don't think it holds true for some of the loaded and collinear types. I haven't yet proved that to my personal satisfaction but I have seen some very interesting test results which seem to support that theory.

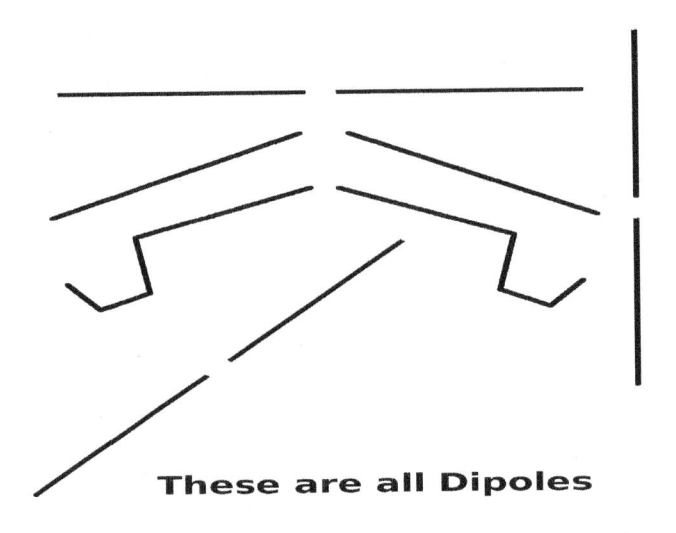

These are all Dipoles

How to Choose the Right Material for your Antenna.

The title of this book is How to Build Effective Wire Antennas. Obviously the material we choose will be wire. But, there are different types of wire and there have been discussions about whether or not you can use certain types of wire for an antenna. My advice is, all wire works as an antenna but you will get better results with copper or aluminium (if you can get it) rather than steel fencing wire or gardening wire.

It has no effect on the ability of the antenna to function if the wire is plastic covered or anything else, except for coaxial cable. Coax will work fine but you have to use the screen as the radiating portion rather than the core, the screen will, um screen the signal.

There is more than wire required to construct a wire antenna. What? Yup, you will need some type of insulating block in the middle of your ½ wave antenna and you will need some type of insulating blocks at the ends to tie it off with.

You can buy special ceramic insulators for that purpose but they are not the only solution. What you can't use is anything which conducts electricity, test with your multi-meter. Some materials don't conduct electricity but do conduct radio frequencies, grey plastic tubing is one of those. This is hard to test for those with limited equipment, stick with the materials I mention and you'll be fine. The other requirement is that it be strong enough to take the strain of supporting the wire.

Materials you can use are plastic, nylon fishing line, nylon rope, Perspex, glass, ceramic and other insulating materials. My usual preference, because I keep changing the antennas, is that 20mm (3/4") black plastic water pipe used for home irrigation. I cut it into approximately 5cm (2") lengths for the ends and 8cm (3 1/4") lengths for the centre. This stuff is strong as. The advantage of using pipe is that you can then string a rope through the middle of it for tying off the ends.

I use 20mm pipe because that's what I have plenty of. If I was buying some specially for an antenna I'd probably go for thinner stuff to make it easier to tie off the rope through, a couple of knots would stop it pulling out.

I also drill holes through the pipe to thread the wire through and sometimes also to thread the tie off rope through when I use nylon fishing line or string to anchor the ends.

I used to overlap the loops through the pipe in case the pipe was cut through by the wire or tie off cord, line or string. It never happened once in 10 years so I don't bother any more.

What thickness wire to use is another serious question I have been asked many times. There is a mathematical formula to determine the adjustments you need to make to the length of the wire based on the thickness of the wire used and I built a spreadsheet to do a what if experiment.

Seriously, it doesn't matter at the HF frequencies. You can't measure 5m (17') of wire accurately enough to get it exactly right so there is nothing to be gained by being that fussy.

What you do need to think of is the strength of the wire. That's far more important than the thickness question and if it's strong enough it will be thick enough.

Remember, if you are using a horizontal antenna the wire needs to be strong enough to be supported in the middle, strung up in the air and tied off at the ends and stay there when the wind blows. Not much of an ask is it?

I usually use the copper wire I recover from the inside of used transformers but if you buy a roll of twin and earth electrical cable or just the twin light cable that works great as well. I measure off the required amount and then split the core out. Leave the plastic insulation on because that adds to the strength on the antenna.

Electrical cable is really good because it's cheap, well insulated, usually copper and thick enough to carry 20A at 240VAC or perhaps 120VAC in your country. Since wire doesn't care about voltage, only current, your antenna will carry 20A at whatever frequency you want to operate on.

To cause this stuff to fail when transmitting you'll have to pump in more than 4,800 watts. Legal power for Hams in most parts of the world is 1,000 watts or less so you will have no problems with this.

That stuff is pretty cheap compared to other stuff. You could also use the thicker automotive cable. The really thin automotive cable isn't strong enough even with the insulation on it.

So you're shopping for wire, nylon rope and plastic pipe. Can you get access to a hardware store? Most of what you need is there.

How to Choose the Right Size Antenna.

We're getting there. You should have already decided on the frequency you want to operate on and worked out how long a ½ wavelength is for that frequency.

Now, because radio waves travel more slowly in wire than they do in air the ½ wavelength of your wire antenna needs to be shorter to resonate at the required frequency. The requirement for a wire antenna is close to 98% of the free space wavelength. Don't get concerned about that, you'll use most of that 2% up just making the loops at each end and the middle.

Now that is only relevant if you can fit the full antenna into your yard space. If you don't have the space we need to have a little chat about how you fit the antenna you want into the space you have.

Shortening antennas is a time honoured method of making stuff fit into small spaces. This is how it works, the area of greatest radiation is the centre of the antenna at the feed point out to about ½ way from the middle to the end. This equates to the middle ¼ wavelength.

A bit of basics, for the antenna to radiate current has to flow in the wire. The centre of the antenna, when there is a gap between the two sections has a impedance, that's like resistance, of 73 ohms. The impedance at the end is very high, not infinity because there is still capacitance to ground. Obviously a lot more current flows in the middle than the ends which means that there is more radiation from the middle than the ends.

If you need to make your antenna shorter to fit your space you can fold it up a bit but don't fold up the middle bits, fold up the end bits.

Its best if you fold them up only just enough to fit, you really want to have as much wire in the air as possible. There's a bit of a trick to this, you can run the antenna in a zig-zag manner in your space, you can even double it back on itself, don't let the doubled back bit get too close to the main antenna because you'll end up with a antenna on a much higher frequency and that isn't what we are trying to achieve.

Here is a sneaky way to fold the ends. Make two insulators for each end, join them with your nylon rope, run your wire antenna into the first one and then loop the wire along the nylon rope so it hangs in loops to the next insulator. Tie off is attached to the second insulator.

Remember that the shorter the antenna is from it's optimum size the less efficient it is and also the smaller it's useful bandwidth.

I haven't mentioned antenna bandwidth before now have I? When you have an antenna cut for a particular frequency it will operate for some frequencies above and below that centre frequency. Edges of the useful frequency is usually considered to be the frequency at which the radiation is 3db down from the the maximum radiation. These are the ½ power points and they define the bandwidth of the antenna. The lower the frequency the closer these ½ power points are to the centre frequency.

The bandwidth for a dipole, that's the proper name for our ½ wavelength antenna, can be calculated to be close to 9% of the centre frequency. This is evenly spread either side of the centre frequency so for our 28Mhz antenna the bandwidth is 2.5Mhz which would cover the whole band for transmitting.

If you are planning something for the 80m band the centre frequency might be 3.5Mhz and the bandwidth would be 0.3Mhz which gives you a usable frequency range of 3.35Mhz to 3.65Mhz. By adjusting your centre frequency you can cover most of the 80m band.

This is mostly relevant for Hams and transmitting, Short-Wave Listeners will find that a receive bandwidth is more related to how much wire you can get in the air than what frequency it is centred on. But, listeners can get stronger signals if they do have an antenna cut to a frequency of interest or a matcher to peak the signals.

How to Choose the Right Type of Matching for your Antenna.

Matching an antenna is actually a misnomer. What you are really doing is matching an antenna system to your radio.

But, what I am going to talk about in this chapter is choosing the type of feeder to attach to your antenna.

Back tracking a bit, if you use a horizontal dipole the feed point is approximately 73 ohms, you can mount a dipole vertically and still feed it from the centre. This will still have the same centre feed impedance.

However it is difficult to set a vertical dipole up for the 80m band as it would be 40m tall, require aircraft warning lights if you are in a landing path etc. It can be done if you really want to but you'll need to consider all sorts of things to do with support, insulation, protection from the very high voltages present at the bottom of the antenna and more. Nup, not worth the effort unless you have a lot of space and can fence the structure off.

If you do a little adjustment and just construct half the antenna vertically and the other half you fold up then the antenna is a lot shorter, you can raise the feed point off the ground a bit and you can put all the bottom half inside an insulating tube, it isn't going to radiate much in any case. This is a ¼ wave antenna.

The signal strengths will be a little down but not much. The bandwidth will be narrower but the kicker is that the impedance will be ½ as well. This means that you'll have to match to 36 ohms not 73 ohms.

There is another issue, they're never ending aren't they? That's because antennas are actually very complex dynamic devices and the maths to describe what they actually do is very complex. I'm trying to avoid the maths and keep it pretty simple.

A dipole is an inherently balanced device because it is the same either side of the feed point. A ¼ wave antenna, or any antenna which is fed at the end is inherently unbalanced.

You need to know this because there are two types of feeder, yep, balanced and unbalanced. You can change this with various matching devices. This means that you can change from balanced to unbalanced and vice-versa.

Choosing the correct type of matching for your antenna is dead easy then, is it a centre fed dipole or not? Centre fed means balanced match, anything else is unbalanced match.

When I get to the feeder choices available I'll explain baluns to you.

How to Construct your Antenna.

Construction is pretty easy. You have already chosen your frequency, worked out how long it should be, got your wire, got your insulating blocks, got some nylon rope and you're ready to construct.

If you are using twin, figure 8 wire or similar you only need to measure a ¼ wavelength for your frequency. Just a quick tip though, if you are limited to the thin automotive type twin wire don't despair, just use the twin each side of the antenna and join both sides of the twin to make one double thickness antenna. This should make it all strong enough but you will need to treat it as single wire for the purposes of measuring the size.

I have a simple rule, all joints must be soldered. The other thing I do is, at the feed point, twitch the wire through each end of the centre insulation block, push a bolt through at each end as well and either use two nuts or a nut and a wing nut. The first nut is to clamp the bolt to the insulation block and the second is to join the feed points of the antenna and the feed line together.

If you are going down this path use tags soldered to the ends of the wires, antenna and feeder, that fit over the bolt rather than the lazy method of just twitching up the wires. Twitching works but your will almost certainly get corrosion happening and this will restrict the incoming signals and cause heat build up when transmitting. Neither of those are helpful.

The feed point is the part that requires most of your attention with a dipole because this is usually the main support point. If it isn't then it will almost certainly be between the support points and have some kind of heavy feeder suspended from it. Either way make it strong, there is nothing worse than your antenna collapsing while attempting to catch a rather tricky contact.

If you are constructing a vertical antenna either ½ wave or ¼ wave then you will need to have some type of supporting structure for the antenna. The support is often an insulating structure but it doesn't have to be. If you have a good look at any of the broadcasting antennas around your location most of them are vertical. The antenna is not the massive tower that is so visible or even the guyed lattice structure that you can see with a little hut at its base. No, the actual antenna is one or more bare copper wires which are strung up the legs of the tower or lattice structure.

If there are multiple conductors then they are all joined at the base and fed from the matching hut. The tower is constructed to be ¼ wave at the frequency of operation. Actually that is a bit of a misnomer. They are constructed to be an electrical ¼ wave but not always a physical one. Do a little maths on the local radio station's frequency and find out why.

All antennas have to be a resonant ½ wave or a multiple of that. This differs from a physical ½ wave as mentioned earlier when I said that the dipole is actually cut to be 98% of the calculated length. Now I don't want to go into the maths of this so just trust me OK?

Assuming that I'm telling you the truth then how does a ¼ wave antenna work? This is where the ground comes in, not the dirt ground but the electrical replacement for the other ¼ wave to make the antenna system appear to be an electrical ½ wave.

Most ¼ wavelength antennas are fed from one end, top or bottom but mostly the bottom because it's easier. Many ¼ wave antennas are constructed of material which is stiff enough to not need any other support to keep it sticking up into the air such as aluminium tubing. Wire tends to fall over and sky hooks really don't exist. See the next chapter for some ideas on how to solve this bit.

How do you get a ¼ wave antenna to look like an electrical ½ wave antenna so it resonates?

You construct an electrical ¼ wavelength at the base of your ¼ wave antenna to act as an electrical ground so you can tune the ¼ wave to resonance. Typically the books all talk about having 120 ¼ wave radials around the base of your ¼ wave vertical but if you had that much space I'd be putting up a dipole instead.

What works is 3-4 radial wires, not necessarily in a straight line out from the base but preferably not overlapping each other either. Bury them just under the ground and pin them down with little wire hoops so they don't catch on anything. They'll work well for you. You can experiment with adding more if you wish, use different lengths as well.

How to Support your Antenna.

Little secret, if you do need to construct a vertical antenna there is a really useful product which you can use to support your wire up to about 6m (10ft) and costs about $30 from your local hardware store. Plastic water pipe, get something around 40-50mm diameter as the smaller ones get a bit floppy in strong winds. These can be guyed by putting eye bolts through them but usually that isn't necessary.

Mounting is easy, drive a metal pipe or steel dropper into the ground leaving about 1m (3ft) above ground, put a bolt through the tubing a bit shorter than the length of the support above ground. It's an easy matter to lift the mast over the support and lower it to the stop bolt. If you use two lengths of this tubing of different sizes you can slide one inside the other, overlap about 1/2m or 18" and put two bolts through to hold them together. Gives you additional height for your vertical or can be used as a supporting mast for your dipole.

If you prefer something more robust then you can buy a metal telescopic TV mast from Tandy, Disk Smith or other source. These are much harder to handle on your own but it can be done. I would suggest though that you make sure that you use some type of anti-seize material on all bolts and the telescopic sections when you erect this as you will want to take it down at some stage.

This is the type of mast I used for many years and I can assure you that it is easier to assemble if you start with it vertical. The way I constructed my antennas with this was to put the base over a post, same as with the plastic pipe, and anchor it to the supporting structure. In my case this was the shed but you may need to guy yours. In that case guy the first section firmly before you do anything else. Take the time to make sure it's vertical now as it is a lot easier than later.

Once that's done, attach a pulley block the the top of the mast and thread it with whatever you are planning to use to pull the antenna up with. This applies even if you are planning a vertical as you will probably want to make some adjustments now or later, make it easier for yourself.

Tie one end of your hauling rope to one of the guy wires or some anchor point you can reach easily later and the other the same. You really don't want to have to pull the mast back down because the rope got away from you.

Attach the next lot of guy wires to the very top of the mast, leaving them slack and lift the end last section up till it hits the stop. Tighten the bolt. Attach the next guy wires to the next section and repeat. What you are doing is pushing the mast up bit by bit starting with the very top section and working your way down the sizes until you run out of mast.

Once it's in position then you can tighten the guy wires. Again, take your time doing this and watch carefully. Many a good mast has been bent by overly enthusiastic helpers who tensioned the guys unevenly.

One mast upright. Well done. Attach the centre of your dipole or the top of your vertical and pull up the hauling rope. I usually attach a cleat to the first guy wire and tie off the hauling rope there to keep everything in place.

If you can't put up a mast of any type then there are still options for you. Attach a nylon fishing line to your TV antenna, the peak of your roof, a tree or any other high structure and attach your antenna to that, string it to another high-ish point and tie it off with fishing line again.

If you have slate or cement tiles you can mount your antenna in your roof space, lie it on your roof and tie it down. On a multiple story house or apartment block you could sneak a wire down the side of the building if you are listening only. I set off the fire alarm in a motel I was staying in once by attempting to tune up my radio with an end-fed wire out the window. Ooops. I went back to listening only.

If you have no other choices you can always run the antenna along a wooden fence, along the ground, through the front hedge, around a window frame etc. You'll have to experiment a bit, or a lot, with lengths etc. to get one of these on the frequency you want but these do work and work quite well. They will never be as good as a full sized antenna at a good height in the air but some listening is better than none. Yes you can transmit through these as well but you do need to make sure no-one can touch these, especially the ends, as they could get quite nasty RF burns from them.

I have heard of people tuning up the railway line outside their home but I would not recommend that. What it does show though is that almost anything will work if you can get the antenna system to resonate on frequency and match your radio input/output.

How to Choose the Right Feeder for your Antenna.

Feeders cause almost as much discussion and misinformation as antennas it seems but there are a couple of things you should be aware of. There are two main types of feeder, balanced and unbalanced. This has nothing to do with their mental state.

Since you now know that there are two types of antennas, balanced and unbalanced you'd think it would be an easy matter to match one with the other but this isn't the case. My preference is to always use a balanced feeder and put it through a matcher so I can adjust the match across multiple bands. This is partly because I am always looking for the optimum match and like to work where the signals are.

If you are only interested in a single band or even just two or three bands then it will probably be easier for you to use an unbalanced feeder, some type of matching at the antenna and switch feeders for each band change.

Unbalanced feeders are usually coaxial cable. This consists of a wire core surrounded by an insulator which is wrapped in a metal braid and then a protective plastic sheath. This is what is most likely between your TV and your TV antenna. This stuff comes in a range of impedances from 30ohm to 93ohm. They can be made with any desired impedance if you want to order sufficient of it.

Balanced feeders can be made by you or you could buy

it from the hardware store. I have used successfully the black 300ohm TV ribbon for 100W transmission without ill effects. The white stuff is really only good for inside use as it seems to absorb water when it rains. You can buy 450ohm balanced feed and this is actually a very good choice when you want to multi band a single dipole. It is rather expensive though. If you are making your own keep the two wires separated by about 30mm to get an impedance around 600ohms.

It is difficult to make and mount but it does work brilliantly. This is really only for the more dedicated amongst you and you can dig up all the proper dimensions yourself. I have done this in the past and used, surprise surprise, black plastic pipe for the separators but I can assure you that cutting 100 separators, drilling a hole in each end and then cutting a slot to the hole, twice, each end for a total of 200 holes and 400 slots and then tying them in place is a tedious process. Worked very well though.

For the rest of us we'll just use the easily obtained and easy to handle 300ohm ribbon or coaxial cable. Unless you already have some experience with feeders I would suggest that you stick with the coax as it is really forgiving, can be run taped against a metal structure with no issues, tied in knots (not recommended) and is strong enough to be self supporting when hung from a dipole tied off at the ends.

Fortunately the most common size is 50ohm coax and this is also a reasonable match to our 73ohm dipole or 36ohm ¼ wave antennas. There is only one problem with this then. The dipole is a balanced antenna and the coax is an unbalanced feed. See the next chapter for solutions.

It's not an issue for the ¼ wave antenna as that is an unbalanced antenna so you can just attach it with the core of the coax to the vertical and the shield to the radials.

One more little trick for you. Cut the feeder coax to ½ wavelength at the same frequency you cut the ¼ wave antenna to. This is an electrical ½ wave and equates to 66% of the physical length for most of the coax you can buy easily. You can easily check if this is correct for the coax you have by looking up the characteristics. There will be letters and numbers on the coax, check those through a web search. What you are looking for is the velocity factor. This will be written as either a percentage or as a decimal. 0.66 is the equivalent of 66% and that is the amount you need to multiply the physical ½ wavelength by to get the electrical ½ wave of the coax you have.

If you do that you can terminate the coax at the antenna and at the radio and expect a reasonably good match across a reasonable portion of the band without doing anything else.

How to Match your Antenna and Feeder.

In the previous chapter I talked about choosing your feeder and you really can match any feeder to any antenna. That's why I suggested that if you are inexperienced with managing feeders and antennas then you probably should go with the coaxial cable because it's dead easy to handle.

There are issues with using coax though. It's an unbalanced feeder and, while it matches fine to an unbalanced ¼ wave antenna it definitely doesn't match nicely with a balanced dipole.

What does it mean to feed a balanced antenna with an unbalanced feeder? What happens is that the antenna currents flow down the outside of the coaxial cable. This is one of the causes of television and radio interference.

It can also cause RF burns from your equipment plus other undesired side effects in your equipment and some extremely odd transmission and reception behaviours. These side effects can also rob you of decent reception or transmission. It is really in your best interests to eliminate this from your antenna system.

This is where I begin talking about baluns. Your TV antenna probably has one, most do. The centre feed point of your TV antenna is on an element which is a special form of dipole, mostly. If you have a closer look at it you should see that the element is folded over. This is called a folded dipole, this antenna stuff has tricky names hasn't it?

The reason the TV antenna manufacturers use a folded dipole is because the bandwidth is doubled and the feed impedance is 4 x greater than a standard dipole.

I'm sure you understand why double the bandwidth is a good thing but why would they like a 4 times increase in feed impedance? Do the maths and you'll see that 4 x 73 = 292 ohms. So that's why that TV ribbon is rated at 300 ohms.

Normal TV coaxial cable is rated at 75ohms and that little balun at the feed point of the dipole has a 4:1 ration so the 300 ohm feed point is matched to a 75 ohm coaxial cable and the balun has converted a balanced antenna to an unbalanced feeder.

It's almost like magic. These baluns are actually pretty easy to construct. You can buy them if you are prepared to pay the dollars but there really is no need. If you are a listener then you can use the standard TV balun to do the matching and attaching to a standard dipole will give you a balanced antenna matched to an unbalanced feed line.

The actual match is ¼ x 73 = 18 ohms which is a bit low and will give you some issues with received signal strength, they'll all be low. How can we lift this back up to 50 ohms at the radio? Assuming that you don't want to play the matching maths game or even play with Smith Charts then lets go with a different solution. Lets change the balun from a 4:1 ratio to a 1:1 ratio.

Sometimes a diagram is required, this is one of those times.

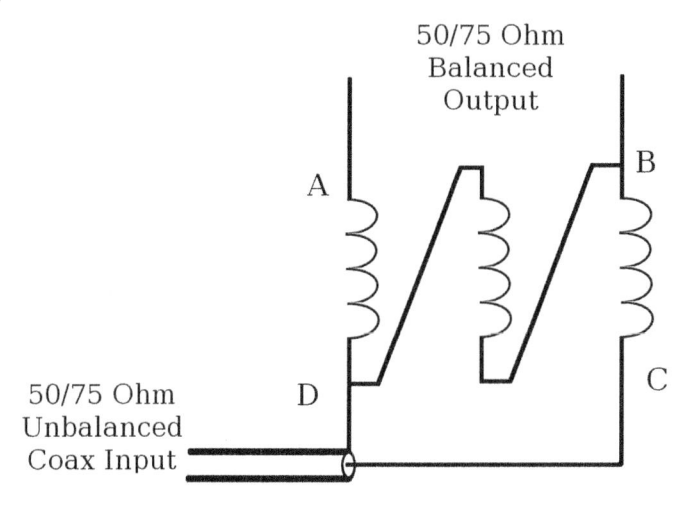

```
                        50/75 Ohm
                        Balanced
                        Output
                              │          │B
                      A  │    │          │
                         ⌇    ⌇          ⌇
                         ⌇    ⌇          ⌇
                         ⌇    ⌇          ⌇
  50/75 Ohm          D   │    │          │C
  Unbalanced             │               │
  Coax Input ════════════╧═══════════════
```

More decisions. How thick should the wire be? How many turns and on what? These things are super simple. Take three pieces of enamelled copper wire about 25cm long (10in), twist together tightly or wrap with tape to bind them closely together, wrap around a piece of ferrite rod of any shape. Keep the turns close together without overlapping. Try using some of the core copper from that electrical cable you built the antenna from.

Do that connection thing according to the diagram and shove that inside a plastic tube so you can seal it off to keep the weather out. You'll need to connect two of these wires, D & C, to a coaxial socket and the other two wires, A & B, to some termination bolts. Solder some tabs on for a more reliable connection and attach the antenna feed points to them.

Hey Hams, this little design is good for 3Mhz to 30Mhz

transmitting with 100W of RF power. If you want to use more you should check into a dummy load while cranking up the power, if it starts to heat up then you've peaked. Get some different ferrite, you'll need to do some research. Don't believe the BS about ferrite rings, they're often more trouble than they are worth.

Just don't try to use this across all the bands with a single dipole, that just won't work. For the band the dipole is designed for it'll work great.

So that covers easy matching from your feed line to your antenna.

How to Match your Feeder to your Radio.

If you use a ½ wavelength, as measured and calculated in Chapter 6, of 50 ohm coaxial cable and you have matched your feeder and antenna then you really don't need to do anything special. Your radio antenna input is probably 50 ohms already.

However, what you will find is that the match is best at a single spot frequency and then rises either side of that point. To keep the match at that low point you'll need what is called an antenna matcher. This is a device which fits between your antenna system and your radio. By adjusting the settings you are able to find a new matching position which keeps the match low and maximises the received signal strength and the transmitted power.

Just a quick myth debunk here. There are people out there who claim to be 'experts' and will tell you that and antenna matcher is another form of dummy load and will rob you of either input or output power. They are totally and completely wrong.

A properly built antenna matcher will improve your antenna system as much or more than any sort of pre or (legal) post amplifier you could make or buy. All transmitting or receiving radios have multiple matching circuits built into them, they couldn't work without them, and none of these 'experts' would ever suggest that you don't need them.

Treat these turkeys with total ignore. If you are one of

these 'experts' then please go do some homework. Accept my apologies for calling you a turkey but go do some real homework before handing out erroneous information.

There are a bunch of different matchers you can buy, most of them pretty inexpensive and easy to use but you can spend really good money for one of these if you like to do that.

It's not necessary though, these too are easy to make with simple tools and they work well. I have one I made years ago which is fitted in ½ an old transistor radio case. I have used this clunker to transmit with 120W PEP SSB from 3.5Mhz to 30Mhz, hams will know these terms listeners don't need to know them.

The basics, a simple coil wrapped around a 10cm (4") length of 20mm diameter (1/2") black plastic irrigation pipe. I wrapped the wire from one end to the other with a little loop over a 5mm rod (use a bit of dowelling) every second turn. Yeah it is fiddly and you'll loose turns and loops and have to start again but stick with it.

I made a hole at the start and pushed the end through then taped it into place. The wire for the coil should be stiff enough for the loops to hold their shape without the dowel in place. When you do have the loops in place push the end through a hole and tape it down. Now you could leave it like this but it's better to run a bead of glue along two sides of the coil to make the mechanical attachment to the plastic stronger.

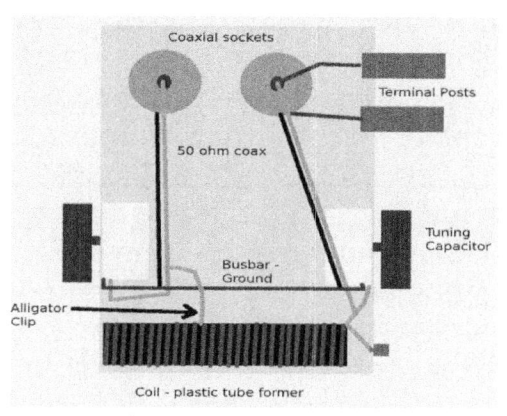

Scrape the enamel off the loops so you can get a good electrical connection. You'll need a variable capacitor, I used the tuning capacitor from the transistor radio. Yes they do work very nicely for transmitting, I wouldn't try 400W though. The thing about these is that they have mylar separating the capacitor plates and therefore are insulated from arcing for reasonable power levels.

These things are cheap as chips. You might like to know how to wire all this up I suppose.

The two coax sockets are mounted through the case. The braid of the coax is ground to both sockets and the bus-bar which is a thickish piece of copper wire. The terminal posts are wired across one of the sockets which connects them through the coax to one end of the coil and the little lug thingy.

The bus-bar is common to one side of both tuning capacitors, the other side is connected to the centre of the coax for both sockets. One of those is connected to a jumper wire and an alligator clip which is used to tap onto the loops on the coil.

The socket with the terminal posts attached is the input and you can attach twin feeder, coaxial cable or end-fed wire. The other socket goes to your radio via a coaxial link cable, bet you didn't guess that one did you.

While you are learning how to tune this take it real easy. Do not transmit through this until you have first peaked the audio by changing tap positions and adjusting the capacitors and then start small.

What I do with a new antenna on this beast is to set the capacitors to their middle position and then try different tapping points until I get maximum noise. I then adjust the capacitors to peak that noise. Once I have that point I'll feed a little power into the system and watch the SWR meter, adjust the capacitors for lowest SWR. Increase the power and tweak as required.

Working it this way you can get an SWR of close to 1.2:1 but anything under 1.5:1 is perfectly fine. If you are not able to get there then try a tap point either side of the one your are using and try again. Walking up and down a band, once you have this matched, generally only requires minor fiddling to keep the match excellent.

If this is a bit daunting, just go buy one and learn to use it. Listeners will find that a good matcher peaked for maximum noise will give excellent results and extend your listening pleasure.

How to Find Contacts with your new System.

Generally the best way to find contacts is to be patient and listen. Work your way up and down the band listening for the tell-tale signs of activity. Some times you'll hear a station tuning up their radio, wait around for a bit and find out if they're going to transmit.

Often tuning signals are not right on the frequency the operator wants to use. They may be able to hear something you can't and don't want to tune up right on top of them. This is good etiquette and you should take notice.

Other times they are just testing some new piece of equipment and have no intention of making any contacts or begin broadcasting just yet. You might like to make a note of the frequency used and check back later. Sometimes they are tuning up for a scheduled contact but are waiting for their contact to call them. It may not happen.

If you do hear a faint signal which you cannot quite make out wait around for a while, there may be others in the group who you will be able to hear and/or the conditions may change sufficiently that you can hear what is being said. Patience is something you'll need a lot of depending on the conditions at the time.

I have had evenings when I couldn't make out anything even though I could hear stations in the background. After several hours and casting about I have given up and headed for bed. Other nights I have had stations all over each other and had to use all the filtering capacity of my radio to isolate stations.

For hams the same rules apply. The main difference is that you should tune up your radio and antenna system on the frequency your antenna and feed line are cut for before looking for contacts as you do not want to be frantically attempting to get the optimum tune when you have just heard that elusive station and he has just said "5 more contacts and I have to shut down". Now you will want to be able to call him back immediately.

You still may not make the contact but you want to give yourself the best possible chance. Just like my favourite contact on 3.55Mhz CW. 5 W to Tokyo, Japan from South East South Australia during the day. Distance over 10,000km – 2,000km/W.

My antenna was a home made wire loop with a balanced feed line and through my antenna matcher above. I can guarantee that the thrill you get with hearing or making contact with other stations using equipment you made yourself is second to none.

After 30 years of experimentation and building I still don't get tired of it. Please take the time to create something for yourself and enjoy this hobby of radio communications more fully.

Printed in Great Britain
by Amazon